GLENVIEW PUBLIC LIBRARY
3 1170 00421 2837
P9-DGT-103

For Hannah and Sophie

Library of Congress Cataloging-in-Publication Data

Schubert, Ingrid, 1953-
[Van mug tot olifant. English]
Amazing animals / Ingrid and Dieter Schubert ; translated and adapted by Leigh Sauerwein.
p. cm.
Summary: Surveys the vast array of animals in the world, focusing on their natural behaviour, habits, and life cycles.
ISBN 1-886910-05-7 (alk. paper)
1. Animals—Miscallanea—Juvenile literature. [1. Animals] I. Schubert, Dieter, 1947- . II. Sauerwein, Leigh. III. Title.
QL49.S26513 1995
591—dc20 95-421

Originally published in Holland under the title *van Mug tot Olifant*
by Lemniscaat b.v. Rotterdam

Printed and bound in Belgium
First American edition

Amazing Animals

Ingrid and Dieter Schubert

Translated and adapted by Leigh Sauerwein

FRONT STREET & LEMNISCAAT
ARDEN, NORTH CAROLINA

Animals both large and small,
Amazing animals, one and all,

are gathered here to say hello.
So turn the page and off we go!

What will pop out of these eggs?
Pointed beaks and scaly legs!

Feathers of many colors,
Armor as tough as nails,
Fur and quills and skin so thin,
And spiraling, slithering scales!

Some animals are swimmers, others like to dive,
and some perch over water or sit carefully where it's dry.

Who glides? Who flies? Who flutters in the air?
And who would rather disappear deep down inside a lair?

There are those who bask in the heat of the sun
while for others the day starts at night.
There are those who are made to leap and to run
while for others to dawdle is right!

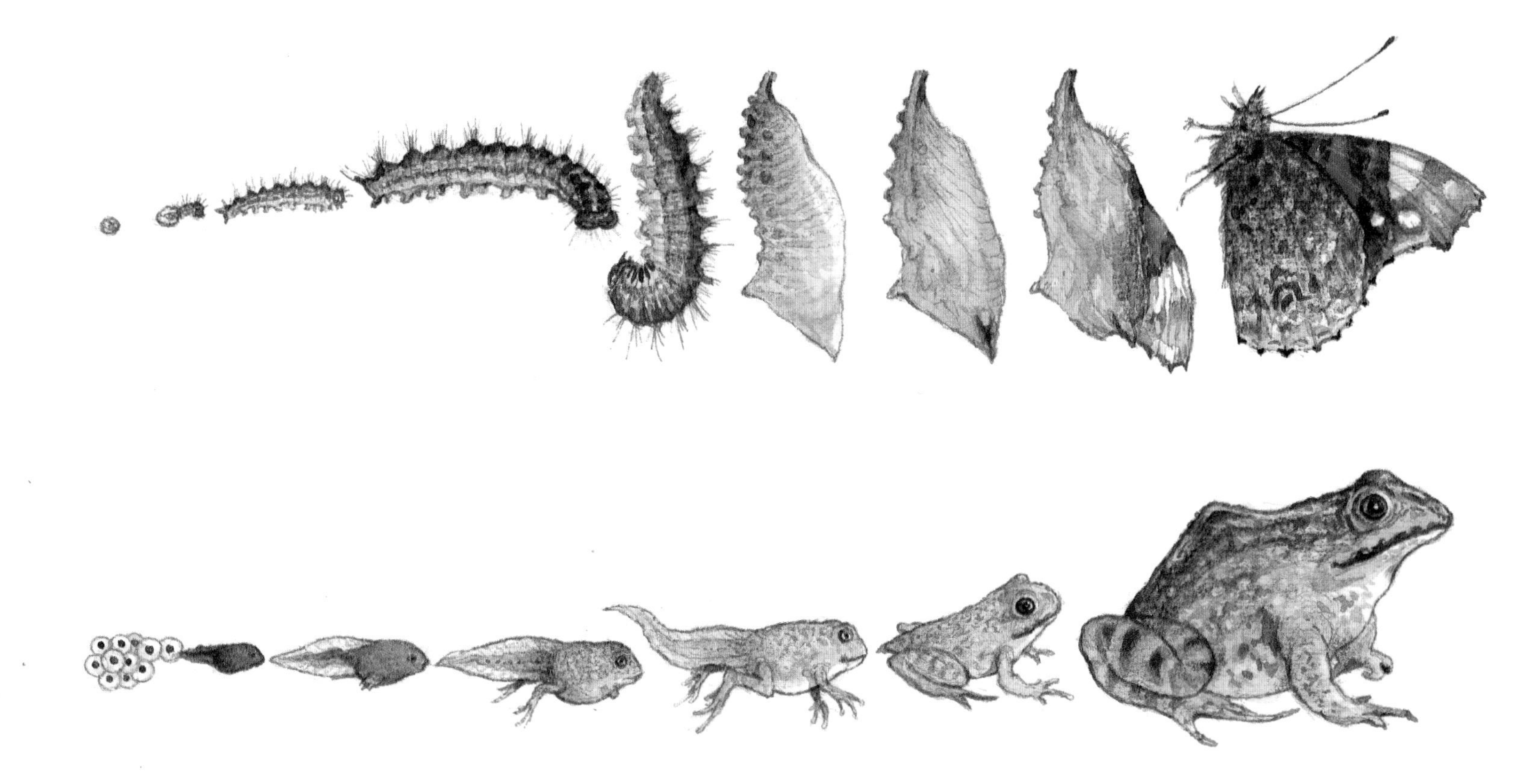

Some even grow up and don't look the same.
What a trick! What a magical feat!

But one can gulp another down,
for living means having to eat!

Insects, insects everywhere!
We count their kinds by millions!
They fly, they leap, they flutter and creep
by billions and by trillions!

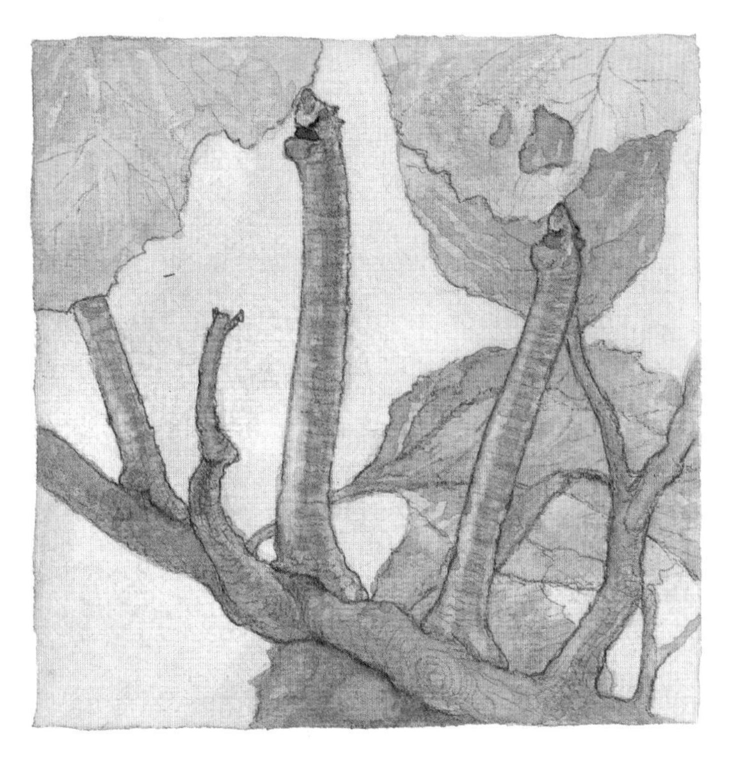

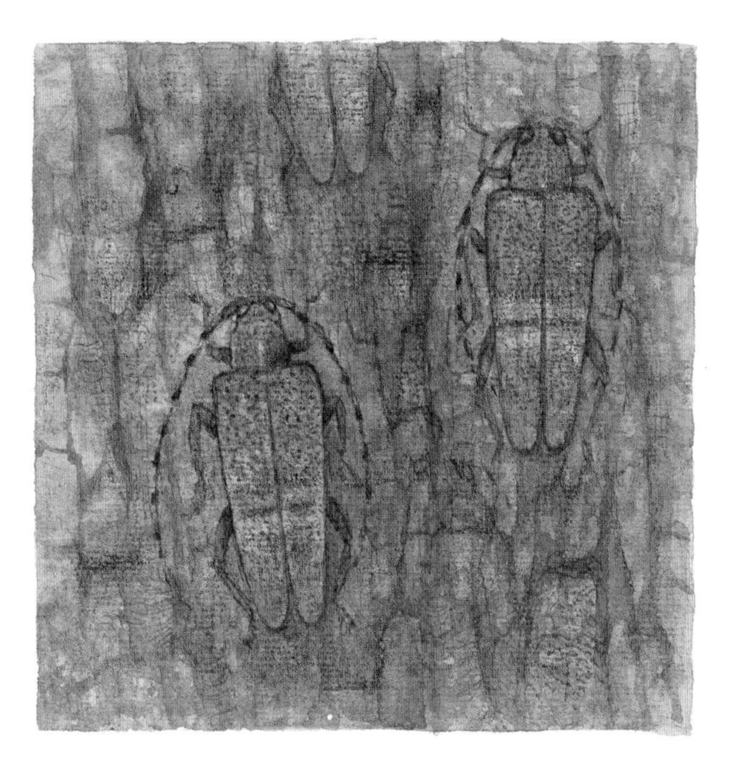

To keep from getting gobbled up,
there's nothing like a good disguise.
Look carefully now and see
if you can find these wise guys!

Rock brown, sea gray, or white as snow,
Who blends in with the icy land
above the water and below?

Some forest folk sit still and hide,
but others like to show

just how noisy they can be,
and how their colors glow.

We're playing dead for safety, that's all!
We would rather roll into a ball!

I shake my quills to be really mean!
We fire a very smelly stream!

When you carry your house on your back, it's easy to go inside.
But sometimes you have to hurry if you want to get home alive!

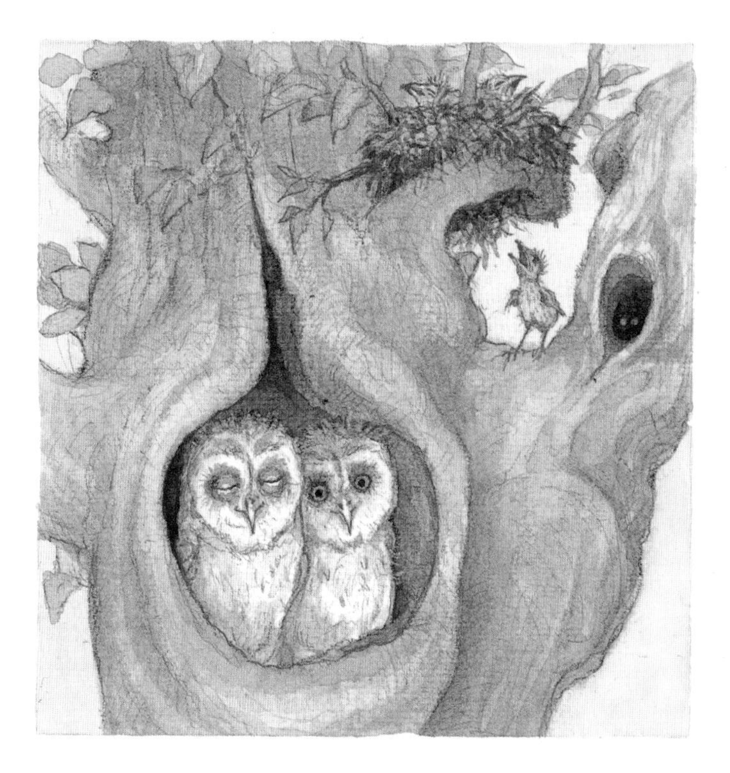

Up high or underground, whose nests are these?
Whose safe and comfortable dens?

Whose baby is whose? That's easy to see!
But if you could choose, which one would you be?

Little cuckoo in the wild
knows he must be an only child.
The mother feeds this sassy brat
and he has gotten very fat!

Greens or grains, meats or bugs,
crunch a leaf, have a slug.
Enjoy yourselves, please do,
but be careful no one gobbles you!

Sometimes it's hard to figure out
how to get your food into your mouth.

It's naptime for this dreamy bunch,
stuffed and sleeping after lunch.

Delicate droppings or dense dung,
every animal's is different, and then some.

Over the side! says Mother. Not in the nest!
Good for you, dear! You're doing your best!

Oh, dung in balls, cry the beetles, our favorite meal!
Let's roll them home, you push, I'll wheel!

Sometimes life is brutally stopped –
one animal catches another and eats it up.

But animals can also grow very old,
and then they look for a private place to die.

Sometimes the family comes to say good-bye.

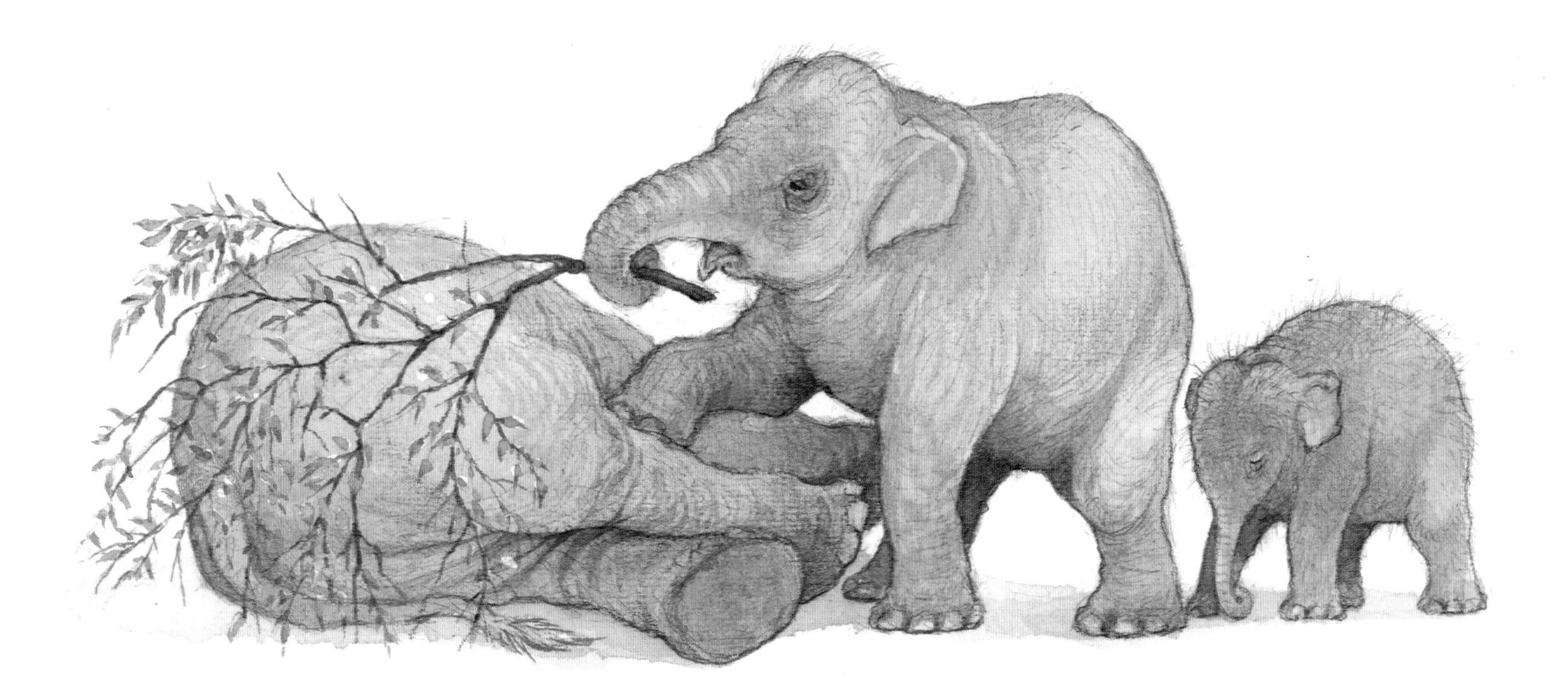

No carcass ever remains untouched.
Busy beaks and teeth will always clean it up.

And a new life begins, as you can see.
What in the world will this animal be?

I have a bill just like a duck
and fur just like a bear,
my babies hatch from eggs,
my home is in a lair.
My paws are palmed for swimming,
and I nurse like any mammal.
Please tell me, don't you think I am
the most amazing animal!